FOOD TRUCK BUSINESS

A Guide to Starting Your Own Food Truck Business and Growing It to Achieve Financial Freedom with Your Passion.

JOHN WEBER

TABLE OF CONTENTS

INTRODUCTION

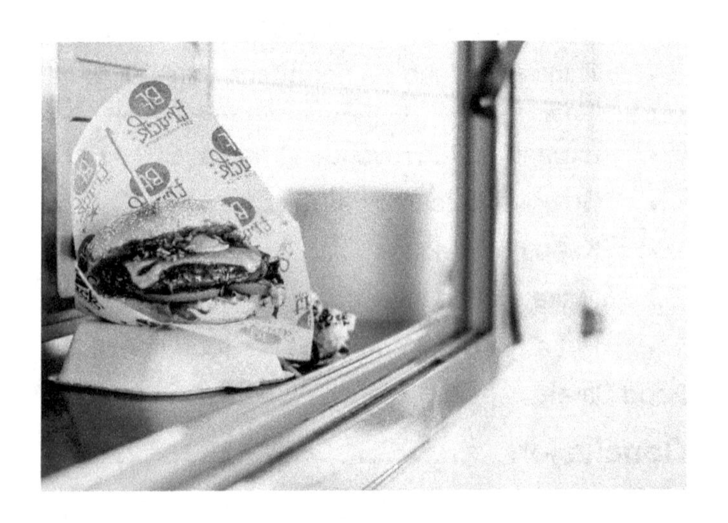

Food trucks in many ways seem like an overnight sensation. Sure, selling food on the street has been around for ages. Ice cream trucks and hotdog stands are nothing new. But the real fun has come from a new twist on an old concept. In recent years, food trucks have gone from selling fries, hotdogs, and ice cream, to becoming something else entirely.

Fresh and exciting paint jobs combine with innovative and creative new food. All types of cuisines are now served off trucks, and their popularity is no doubt expanding. A lot of people, myself included, have looked to this new market as a possible business venture for themselves.

What's not to love? You get to set your own hours, prices, make your own food, control what you sell, and reap the rewards; all while having fun serving food on the street. Whether you're a chef of many years or someone who wants to take a crack at being an entrepreneur, food trucks are a great way to run a kitchen and manage a business.

Unfortunately, it's not always as easy and profitable as it may appear. If you've seen the movie Chef (you should see it, it was great), it may seem like long lines are just waiting for you to pull up and open your truck window. Within a few weeks, Chef's truck has people lining up around the block just to get a taste of what he's cooking. After the first few weeks on my truck, I can tell you that the lines weren't quite that long (nowhere near).

Owning and operating a food truck business for the past two years has taught me a lot of lessons, most of which aren't necessarily skills you've already learned in the kitchen or a business course. This book is meant to prevent current and aspiring food truck owners from making simple yet costly mistakes, as well as showing how it's easy to make a profit if you follow a few simple guidelines and stay on top of certain key details.

It is not meant to be a book on how to make an awesome double cheeseburger, or how to master double-entry accounting. In short, this book's intention is not to teach you how to cook awesome food, or how to run a small business. Those are skills that require a lot more than simply reading this book or any particular book for that matter.

What this book is meant to teach you is how to run your food truck business more successfully, while avoiding some of the mistakes that a lot of people make when first starting out. All the methods listed are meant to help you:

- Get repeat business
- Keep losses to a minimum

- Avoid first-timer mistakes
- Satisfy customers
- Maximize profits

A lot of tips listed are things that may seem very common sense but are nonetheless important. In a lot of instances, the most common-sense things can slip our mind, and leave us with unhappy customers or a bottom line in the red, also known as NO PROFIT.

It's easy to just focus on what you love to do: cook food, or run a business. They are the smaller things, however, that can often mean the difference between making and losing money, and if you're just starting, it will be crucial to not only be consistently making money but to avoid those big losses that can cost you in the long run.

By the end of this book, you'll be able to:

- Avoid costly and embarrassing mistakes that often arise on a food truck
- Increase your chances of long-term street service success
- Distinguish yourself from other trucks

All of these tips I have learned are critical to having success on the street and are things that I wish I could have avoided or taken advantage of the first time around. That is why

I wrote this book; so that you could learn from my mistakes and avoid them your first time around, increasing your odds of immediate success in the food truck business, and paving the way for a long-lasting, profitable business.

Why do I feel qualified to write a book on running successful food trucks? Well, for one, I've been at it for two years. They've been two trying and difficult years in a lot of ways, but also some of the most fun I've had both in cooking and in working. For me, in many circumstances, I was simply winging it and going with what I thought was best at that time.

Looking back on it, there were a lot of benefits to figuring things out on my own, but there were also many avoidable costs. Learn from a few of my failures and successes, and I believe this book will prove to be worth much more than what it costs.

So, without further delay, let's get to trucking.

Let's start with opening up for the day…

CHAPTER 1. IS THE TRUCKER'S LIFE FOR ME?

Starting any business takes time. Even though, so long as you make the effort to start a business, you ought not to fear about any capacity faults that can arise when running the business itself. That notion additionally applies to folks who count on getting into the mobile food industry, although don't anticipate getting concerned in the enterprise... until they're approximate to make contributions.

Questions to Ask Yourself

Though, before you determine what kind of meal truck you are planning to open, it's essential to invite yourself questions about what you could assume out of the business.
Those questions may include:

- What's my passion for food?
- What type of food do I want to serve?
- What type of food may be missing from my community?
- Are there other food trucks with the same concepts/food products as mine?
- How can I differentiate my brand from other brands?
- How much money will I have to initially invest in this business?

- Where am I going to get the food that I'm planning to serve in my business?
- How much do I need to invest in those products?

Asking questions clearly allows put off the uncertainty related to beginning a food truck enterprise. In fact, for numerous people, understanding greater approximately what to expect from a meals truck business makes them greater relaxed with beginning the enterprise inside the first location, in view that they do know what to anticipate after doing the studies. Once you analyze extra approximately what to anticipate, there are different elements that you have to take into consideration, especially while you're, you guessed it, consider beginning a meals truck enterprise. Several the ones factors are a quite imperative part of the enterprise itself, making it quite vital to take note of them.

The Most Integral Factors of Starting a Food Truck Business

The most essential factors of beginning a meals truck company are quite straightforward when you recall it. They're many identical factors concerned with starting different businesses because you quite load want to attend to the one's matters earlier than taking off on the road for the primary time!

Let's take a look at the one's elements, as paraphrased from a mobile food agency useful resource:

Legalities

Understandingly, you can't precisely run a meals truck business without looking for a few forms of permit from your city and country jurisdiction. Having a allow it without a doubt is especially designed to cowl a mobile agency looks after, nicely, the legalities associated with strolling such an enterprise and operating the automobile interior your city.
At remarkable, a permit safely covers strolling most elements of your mobile food truck, besides you want greater criminal permissions from your city or county jurisdiction.

Vehicles

Manifestly, you want to discover a truck to conduct and deliver your business everywhere you want it to adventure. Now, you ought to spend plenty of cash on a food truck. Many food-prepared trucks and/or trailers can fee as low as $1,500 and as high as $75,000, depending on the type of truck you can plan to ultimately buy.

In the end, you need to head for a truck or food-organized vehicle it's miles big enough to house the company and remaining some time.

Logo

If you need your food truck to capture, you will need to set up a brand. A brand is what enables capability customers proper now understand a business, particularly if that business is already quite well-known. As a food truck proprietor, you are going to want human beings to recollect the decision of your business and the ingredients you eat, so begin brainstorming names and thoughts that might foster what you envision your enterprise being.

Names also are just one part of setting up a business; you also have to give you an ability menu, decide out what additives and meals to buy, and put together whatever you need to address earlier than even getting the legal paperwork out of the way.

Financing

Arguably the maximum important issue of starting a meal truck. Why? It's far in particular due to the fact you absolutely cannot do loads with little to no financing. You cannot even purchase the truck! So, in advance, once you get into the entire planning issue, search for capacity non-public groups.

You could have the exceptional desirable fortune looking for a private investor who can be interested in backing your food truck business, although that rarely occurs till an investor definitely takes location to love your imaginative and prescient and your food. you could typically head on your financial institution to get a loan, specifically in case you qualify for any of their financing options—naturally, you could typically be searching for help from an opportunity monetary corporation if you do now not want to apply to a bank.

Financing a food truck normally consists of protective expenses for the truck, branding, system, food and its associated additives, point of sale (POS) or credit score card systems, protection precautions, and every so often, employees.

So, as you could see, several factors bypass into starting your private food truck business. at the identical time as we surely protected the ones particular factors in brief right here, we will evaluate those standards and boom on them a hint greater presently. Even though, for now, allows taking a look at your alternatives close to beginning your very own meals truck business.

CHAPTER 2. WHAT IS THE DIFFERENCE COMPARED TO OPENING A RESTAURANT?

The last things we'll cover before we get into the "meat and potatoes" of starting your own food truck business are the pros and cons of having a mobile food concession versus a brick-and-mortar restaurant.

Having been involved in both types of ventures, I prefer to look at these as "benefits" and "challenges," because both business models can be simultaneously successful and complicated. There is no "better" option, and there is certainly not an "easy" option. In fact, there are quite a few similarities throughout the process, especially in the food sourcing, storage, preparation, and staffing areas. Many restaurants successfully make the switch from brick-and-mortar to mobile and back again, and some easily dabble in both enterprises. Therefore, let's look at some of the key points in which they differ, focusing on the specific benefits and challenges of running a food truck.

Location

You've probably heard the phrase "location, location, location," about everything from opening a business of any kind, to purchasing a home, to marketing and beyond. A cactus might look cool on your coffee table, but if it doesn't get sunlight and warmth, it's not going to do well. Location matters.

Some say the biggest advantage of having a food truck is the flexibility of location. "You can go anywhere!" they say. "You can even hit multiple places a day!"

Traditional restaurants rely on customers coming to their location for the bulk of their business, with the occasional foray into catering, take-out, or delivery as appropriate for their business model. Food trucks draw a crowd wherever they park. In many cases, the customer base is built into the location.

For example, you may have seen a food truck parked at a brewery that otherwise doesn't serve food. This creates a symbiotic relationship between the brewery and the food truck. Some customers come for the beer and stay for the food, others come for the food and stay for the beer.

This seems like the perfect plan, right? But some key unpredictable factors go into making this plan work exactly as anticipated. First, the brewery needs patrons. Maybe, the brewery is hosting live music tonight and half the town has turned out to see the performer, drink the tasty libations, dance, have fun, and specifically eat your delicious food. Everyone wins. But what happens when the bar down the street has the live music, it's raining, and the brewery has run out of their most popular draft? In this instance, it's just another night for everyone involved.

In theory, a food truck could pack up and go wherever people are gathered, but it doesn't always work like that. Setting a food truck up is generally a lot more involved than parking and opening the service window, though admittedly, some specific trucks have been designed with that exact business model.

Therefore, while some trucks appear at multiple locations in a day, it's not the norm. And while you can set up a food truck just about anywhere, there's a significant amount of thought and planning that goes into the process. You need to ensure that foot and vehicular traffic isn't impeded, that all the equipment can operate, and that you maximize the experience for everyone involved, from the owner of the food truck to the guy who just happened to be walking by and smelled something delicious.

Still, the ability to move as needed is an obvious overall benefit to the food truck operator. Even if you have a contract to work a fair for a few weeks, at the end of the contract, you can wander where the wind takes you.

Having this flexibility allows food truck vendors to make more informed decisions about the gigs they choose. Remember, if your business model is based on going to the crowd, you will need to research the venues at which you park and the gigs you accept to ensure that the traffic will be worth the cost of operation. While you have the benefit of moving your restaurant from place to place, once you're set up, you're attached to that spot for a significant amount of time.

Travel

Traveling itself can be both a benefit and a challenge. While having the wheels to spin and the roads to explore are very significant benefits, remember that traveling has a price. You will need to keep your food truck well maintained and in good mechanical order. You also need to be prepared for the occasional breakdown, tire blowout, or traffic accident. Everyone has the best intention of driving cautiously and carefully, doing everything to keep their vehicle functioning perfectly, but accidents do happen. You'll need to fill up on gas when it runs low. Tires will need to be replaced. Oil changes will be needed, not just for the fryer, but for the engine as well.

Everything you do for your daily driver car; you'll need to do for your food truck.

If you choose a concession trailer instead of a food truck or van, you'll also need a vehicle to pull it. That means maintenance for both the trailer and the vehicle, at least in making sure that the wheels, hitch, connectors, blinkers, lights, axles, and wiring are working appropriately when the trailer is on the road.

The amount of traveling you do directly impact the amount of maintenance your vehicle or vehicles will require. Before you plan to travel all across your state, get a feel for gas prices and the maintenance requirements of your truck or trailer. If you want to do a lot of traveling, rather than taking advantage of the local market, be prepared to work the related expenses into your business plans.

This may seem obvious, but the food also has to travel. The general rules of food-service safety indicate that cold ingredients must be kept below 40 degrees Fahrenheit, and hot food should be kept above 140 degrees Fahrenheit. We'll get into food safety in far more details later, but temperatures between 40-140 degrees Fahrenheit are what the United States Department of Agriculture considers “The Danger Zone” in which bacteria are most likely to spread in food items. That means that if you have far to travel, you have to be very conscious about food storage.

Likewise, utensils and serve ware are also along for the ride. Just as the chuckwagons of yore needed plenty of cabinets, drawers, and bins for storing everything that traveled the trails, your food truck needs to be equipped to securely hold everything that’s hitting the road with you. You don’t want to arrive at a gig with a truck full of scattered, dirty, and/or damaged serving containers. Without serving gear, you may not be able to actually work the gig as intended.

Weather

You may have noticed one particular factor in the earlier brewery example that no one can control; weather. Food trucks may provide shelter for the workers inside, but your customers will need to brave the elements to buy your wares.

Snow and rain will certainly deter customers from standing in line for a significant amount of time, no matter how delicious your food is. Thunderstorms and high winds can be downright dangerous since food trucks are a combination of metal and movable pieces that can easily blow around and cause damage to someone or something. Weather that's bad enough can prevent traveling altogether.

Too much sun or heat and humidity can be an obstacle as well, but in this case, more for the people inside the truck. Food trucks require ventilation, but anyone who has worked in a commercial kitchen will tell you it gets very hot, quickly. If you and your staff are running orders very fast on a hot day, surrounded by hot grills and fryers, the chances for heat exhaustion or heat stroke increase very quickly.

Some things can be done to mitigate against the weather, such as outdoor heaters along the queuing area when it's cold, water misting fans when it's hot, or a tent or awning over the service window for any weather condition. These elements are often dependent on what the venue can accommodate or allow, however. Some cities have regulations against where tents can be put up, and outdoor heaters may be against city ordinances as well.

Weather and the elements are indeed huge factors in the food truck industry. And because the tools and equipment available to combat them can be expensive to purchase and run, many food trucks operate on a specific seasonal schedule.

Responsibility

As a food truck owner, you call the shots, and for most people, that's a huge benefit. If you only want to go out every other Saturday, that is your call. If you want to quit your day job and set up at a new place every day, it is certainly your prerogative.

In fact, you might feel like you're starting ahead of the curve by investing in a food truck instead of a brick-and-mortar restaurant. There's no rent or mortgage to pay, and no landlord to satisfy. You don't have to buy a dining room's worth of expensive chairs and tables or hire a decorator to do your colors and create an ambiance. The overhead is initially lower, because you buy a vehicle, insert the equipment, and that's it, right?

Financially, there are risks and rewards every time you put the truck in park and fire up the propane. There's food cost, operating expenses, and travel expenses to balance with profit. You can only make money if you take the truck out and sell food. It's a classic "spend money to make money" scenario. And if you have a slow night, or that package of food containers spills during transit, or you drop a head of lettuce before you can chop it into individual salads, you've spent money that you're not getting back.

One sliver of a silver lining here, though, is that much of the money you take in is going right back to you and your business. Food trucks and mobile kitchens of all kinds have a far smaller staff than a standard restaurant. Even if you have a food runner, you do not have a full-service staff. There are no cleaners or bussers. Due to lack of space, it would be impossible to have various stations, such as the line cook, prep cook, sous chef, expo, and kitchen manager. In a food truck, when the owner is also the head chef and working alongside staff, this means a dramatic reduction of payroll costs. And, if it's a solo operation, there are none. Minimal staff means minimal payout at the end of the night.

CHAPTER 3. HOW TO CHOOSE THE RIGHT FOOD TRUCK?

A vast cost disparity separates the simple food cart (about $2,000) and a full-service food truck ($100,000). The reason for this price difference is that full-service food trucks have specialized equipment for the preparing of food, as well as selling it. With the food cart, you're equipped to sell food, but you're not equipped to compare it. Your gross profit margin will thus usually be lower with a food cart because the food you purchased wholesale is already ready to retail. Full-service food trucks give the owners space and utilities to craft their own inventive cuisine fresh and on-sight, thus their profit margins are higher and their food is fresher and better prepared, allowing them to charge more for it.

Your selection of a food truck should derive from your step 1 "concept," specifically the type of food you're preparing and the equipment and space you'll need to prepare it. The cost of your food truck will be your biggest expense, so you need to take as much time as you need thinking through your preferences, your options, and the available units on the market. For example, if you're going to be selling cupcakes primarily and doing a lot of your baking off-site, then maybe you don't need much space for cooking and preparation, but

you probably should find a truck with a lot of available display area, so you can show off your enticing baked goods.
Consider buying a food truck at a lower price range of $ 40-60K and remodeling it to fit your concept. This is not as expensive as you might think and will alleviate some of the financial risks. To aid in this specific process and the food truck buying process in general, you will fare better if you can acquire the services of a competent food truck designer. Usually, companies that sell food trucks will have someone on staff who can help you with this task.

The essentials of designing your food truck include:

- Making a list of all the equipment you're going to need for your particular concept.
- Obtaining the dimensions of your needed equipment.
- Determining how much overall space you will need in your food truck, both to accommodate the needed equipment and to give you and your employees enough room to operate inside the truck.

- Creating a blueprint using a sheet of graph paper, or a software equivalent. The blueprint will show where your equipment is placed, your display area, where your employees will prepare the food, and where they'll interact with customers.

After you get down these basic requirements for your food truck design, an experienced food truck designer should help you polish off the details. As a warning, make sure the motivations of your designer are on the up-and-up, especially if she works for the company selling you the food truck. Your design assistant is not there to convince you to buy a larger truck or equipment that you don't need but to be a sounding board and fellow brain-stormer as you finalize your design requirements.

Your vehicle designer—though she may not be familiar with the specific state and local laws to which you're obliged—should be able to assist you in meeting basic health standards. Before you open for business, your vehicle will likely be inspected and certain legal and health-related standards will be verified.

Possible verification points include:

- Proof that you own the vehicle and that it is licensed.
- Food purchase and storage records that are up to date and accurately kept.
- Verification that your food truck is supported by an approved and adequately equipped depot for water, general food supply, cleaning, and waste disposal.

Health department food vehicle inspections are conducted usually at once-a-year intervals and sometimes at random. Food equipment, water supplies, and sinks are all checked during these inspections. Inspections may carry over into the kitchens or garages, where the food truck is sheltered to ensure that the area is conducive to housing the truck and that fire codes are being met.

In addition to the unique inspection and licensure requirements that come from the food truck's foodservice capacity, the truck is also subject to the standard regulations that govern every on-road vehicle, such as up-to-date registration and insurance. The motor vehicle department can clarify for you what your food truck will require to use the roadways. Your vehicle will probably be required to have commercial plates, though you will perhaps not be required to obtain a commercial driver's license to operate it so long as it's under 26,000 pounds.

When choosing where to park your food truck, you will again be subject to local ordinances.

Some common local regulations include:

- Food trucks must sell only to the sidewalk side of the street. This prevents customers from forming a line on the street and getting in the way of traffic.
- Food trucks must be x feet away from a bathroom at all times. Southern California, for instance, has a rule stating that food trucks setting up for business must be within 200 feet of a bathroom if the truck is going to be parked for more than an hour.
- Food trucks cannot park within x feet of a brick-and-mortar restaurant. The tension between food trucks and traditional restaurants has gotten so high-strung in recent years that many cities have taken the role of peacekeeping into their own hands.

They've done this by making compromise ordinances mandating that food trucks stay 500 to 600 feet away from established brick-and-mortar restaurants. In the event, these laws exist in your area of operations, make sure you get clarification on what exactly constitutes a "restaurant." Sometimes convenient stores and other non-sit-down food vendors technically count as a restaurant and are protected by the law from your competitive presence.

- Food trucks must be parked at commissaries when they are non-operational. Some cities are very concerned about where your food truck is going to be docked at night. Commissaries are used often as space where food truck operators can access some raw materials and prepare for their day. Even, if you don't need the food-related service that a commissary provides, you still may need to rely on the commissary or another service for a space to park your vehicle when it's not in use.

- Food trucks may only be parked on private property. Certain municipalities have decided that food trucks in public space are just more trouble than they're worth. If you're facing these types of restrictions, then you'll need to put extra work into cultivating relationships with local businesses, office parks, etc. For a food truck, a lunchtime position in an office park, especially one that's far away from any restaurants, is an incredibly lucrative proposition.

CHAPTER 4. FOOD TRUCK BUSINESS PLAN

Many newcomers to the food truck world will often feel overwhelmed by many of the regulations and formalities that may arise. One of these formalities is the creation of your business plan. This document is essential to obtaining funding and is a guideline for the first few years of operation. Several segments should be included in a business plan starting with the Executive Summary.

Executive Summary

In order to inform potential investors and lenders as to why your food truck business will be successful, it's important to develop the proper executive summary. Make sure to provide information about how your business ideas were developed, as well as future plans for your food truck. An effective executive summary should only be one or two pages, highlighting the core concepts and operational procedures for your food truck.

When writing an executive summary, make sure to explain the general purpose in an overview. Include your name, company name, area of expertise or niche, and other basic information such as products and services you plan to offer. Include some menu items, additional services (like catering and special events for example) and explain how you will provide these services to your customers.

An executive summary should also include a price analysis to determine how much startup money you'll need to raise with as well as how those funds will be used within the startup and operation of your mobile food business. In addition to financing, a mission statement is also important. It will tell investors and lenders why you want to start a food truck business in the first place. Again, it might be informative to include menu items or how and why your community will benefit from the formation and operation of your particular food truck business.

Within the executive summary, make sure to add any past experiences that have brought you to the decision of opening your own food truck. List some of the direct and indirect skills you have that will be an asset to your success. A general idea of financial forecasts like costs and profit models is pertinent in the executive summary.

Company Description

The next step is to come up with a company description. Similar to the executive summary, the company description works as a mission statement, laying out the foundation of the company, such as the purpose, daily operational plans, and future expansion goals.

Your company description should also include brief information of some of the closest competitors in your area—this can include other food trucks, food carts as well as brick-and-mortar restaurants. It's a good idea to describe in detail how your mobile business will distinguish itself and stand out from the competition. An example of some of the items, you can list may include your focus on quality, nutrition, fresh ingredients, locations, etc.

Following the company description is a detailed market analysis that examines the state of competition in your area and how you fit in.

Market Analysis

Within the market analysis, lenders and investors will learn how well you have researched the mobile food industry in your area and what other competitors are out there trying to take a slice of the same pie. In addition to general highlights, it's important to include any additional marketing research.

Make sure to include a general description of the mobile food industry, including current developments and potential characteristics of the business model. Depending on your market, make sure to include local and national trends for all types of mobile cuisine most similar to your niche and style.

The market analysis should also include specific information on your target market. Essentially, this is the group of customers that will most likely stop and buy your food. It's important to be able to identify your ideal customer because if you focus too much time and resources on the wrong customer, it could hurt your business in the long run. So, it's best to be realistic, and target the most likely customer for the type of food and reputation that is the essence of your business.

Defining a Target Market

You need to be able to identify and understand the characteristics and mindset within your target market and make sure to include that information in your market analysis. Understanding the needs of your customers can be listed in terms of the type of food that sells and nutritional needs for example. Most likely, you'll find that there are needs that are not being met in your city or region. This gives you the prime advantage of making your business a necessity in the area.

In addition to characteristics, you should analyze the size of your market. You'll probably find that your best customers will come from a select group of individuals. If you find another truck, cart, or traditional business similar to yours, do some investigative research to try to establish what their target audience consists of. This vital information can help you compete and even differentiate yourself as a leader in the region you will be operating your truck. It may help sway customers away from competing trucks.

With the target market information in hand, you can begin to explain and describe your pricing structure and determine the most cost-effective tactics to be profitable. If this step seems overwhelming, consider hiring a professional to do your analysis for you. This may be money well spent so you have better-researched analytics before your hard costs start adding up. An option could be to research using local trade magazines or news articles published in your area.

Once all the pertinent information has been gathered, the market analysis will also need to explain how you will reach your target market. This could either include social media or other types of advertising. Consider the various types, depending on your target audience. Radio and television reach the masses but are not often used in the food truck industry. Social media is often the most effective and is widely used in the mobile food industry.

When comparing which marketing and advertising strategy work best for your mobile food business, take into consideration your market analysis results. Include those methods that you have already used within your market analysis, listing specific examples for investors. This may include potential customer acquisition, the number of direct competitors in the area, or even the price of ingredients.

Finally, when composing your market analysis, the last step is to evaluate the competition within the area. This includes other food trucks as well as brick-and-mortar locations that sell food similar to yours. Research your competitor's menus within your region, making sure to include all the pertinent information, including specific locations, prices, and more.

Management & Organization

Here we will describe precise methods for the day-to-day operations of your food truck, including some employees, necessary preparation work, organizational structure, and more. Make sure to also include how profits are to be broken up among partners as well as each employee's role and duties.

Many partners often find it difficult to confer how profits should be split, but it is extremely important to have this information in writing. That way, everything is spelled out to avoid conflicts and disagreements later. This segment of the business plan also defines everyone's roles in a way to make sure no job is forgotten. Finally, describe the background and general information on all team members.

Products & Services

This is perhaps the most exciting segment because food truck owners will be able to include their cooking methods and new recipes. Again, make sure to explain your truck's ability to fulfill your customer's needs. Include the names of your dishes and any other specialty items or unique services.

Describe your ability to bring revolutionary areas to an area and explain in detail the qualities that separate you from your competition. Describe how your food truck will create raving lifelong customers and how you will bring future plans to life, including how you will book events or gigs for example.

Sales & Marketing

After describing in detail why customers will choose your truck over the competition, use this segment to explain how they will first learn about your business. This segment needs to include how you will attract new customers to your food truck and how you will keep the regulars satisfied and coming back for more!

Whether you decide to give away samples, use social media, send out a press release, or pay for advertisements, remember to think about long-term solutions for growth. Because the industry is so competitive, marketing is just as important as your famous brand and recipes. This needs constant monitoring so you can make changes as the market changes.

In terms of sales, think about how you're going to keep afloat and continue to increase your profit margin. Break this segment down to include competitive menu prices and goals.

Include workdays and hours you'll be in operation, and remember to take into account possible problems like harsh weather conditions, employee issues, or vehicle breakdowns.

Funding & Investors

This segment is crucial for those looking for outside investors. Investments can be necessary for the truck itself or help with acquiring licenses from the city. This segment needs to include specific numbers on how you will use your funds.

In addition to funding needs, make sure to include how and when these loans will be paid back. Believe it or not, investors want to know when they will get their money back! With bank options, the bank will often set this type of information up for you, but it's a good idea to include those figures here. Be clear on money needs and budget correctly for future success.

Also, when looking for an investor, consider including an appendix at the end of the business plan for those unfamiliar with your business to find their way through your business plan with ease. The appendix can include additional documents like license and permit copies once you've acquired them.

Financial Projections

This segment will give additional details about profits and future plans. Use current sales plans and figures to create monthly or quarterly estimations of profit, including all costs and potential areas of loss due to truck issues or outside factors beyond your control. You have to expect the unexpected in this type of business!

In financial projections, many of the facts will feel like pure speculation, but it's always better to plan ahead and at least have a well-researched estimation of your financial future. Once again, be clear and specific to factor all costs.

CHAPTER 5. FINANCING

You must have sufficient capital to not only start your business but also to maintain it. It isn't until you project your costs, you will know if you currently have enough funds out of pocket to make your business idea happens.

There are several options if you do not have sufficient capital out of pocket. Obviously, the best choice would be to save until you can pay for most of your startup and operational expenses out of pocket to eliminate a lot of unnecessary debts in the beginning. But we all know, especially since the economic decline in 2007/2008 that is very difficult these days.

There are various types of financing and we will talk about the most common ones:

Debt Financing

Debt financing is obtaining capital such as loans without giving up equity or ownership in the business to obtain it. Debt financing can come from financial institutions, lending institutions, and non-traditional lenders.

The lending requirements vary among lending institutions. They do not give you money just because you are starting a business or because you are "Mr. Nice Guy or Girl."

They are a business themselves and one of the ways they can earn revenue and stay in business is a client's ability to pay back the capital they borrowed plus the agreed-upon interest.

Because lending institutions are dependent on a client's ability to pay, they implement qualification thresholds to eliminate a lot of the risks in lending. Typical lending considerations although vary from place to place, all have typical minimums between all of them which are dependent on the capital amount you are requesting and other requirements. They will evaluate you based on the financial documents you submit as part of your loan application. The typical documents they will assess are:

- ***Satisfactory FICO score and credit history***

Scores they typically have accepted were at minimums of 680 to 720 or higher. Some institutions have accepted lower credit scores of around 650 but it was more likely due to the individuals, other financial documents surpassed the institution's expectations.

- ***3 Years of most recent tax returnS***

This is how they evaluate your income level interests to determine that if you had to return to the workforce, would your expected salary be sufficient to pay back a loan for the capital amount you want to request.

- ***Collateral***

The higher the capital amount you are requesting, the more they will expect you to have assets that are a 1-to-1 match in value. For example, if you are seeking $200K or more in the capital, they will definitely expect you to have home equity of an equal amount. The purpose is the financial institutions want their money back ASAP and if you default, they want to be able to liquidate the assets you have quickly to get it.

- ***Industry experience and business plan***

They typically ask for your resume to show your industry experience as well as your business plan. These two sets of documents provide them with the opportunity to see if you have experience in the industry you would like to start a business in and if you have a strong plan in place to make this business successful.

- ***Can contribute at least 25-30% out-of-pocket***

This is a set-in-stone standard for most financial institutions. This also helps them see that you have faith in your business idea.

- ***Industry and growth trends***

They will look into the industry and see if it is growing, declining, or a stagnant market. Sometimes the approval process isn't about you and your talents and plans, sometimes they will not approve someone because the industry they are getting into is declining and that scares the financial institutions.

- ***A written explanation of the amount seeking and a breakdown of the associated costs***

Financial institutions will often request a written statement of the amount you are seeking, plus a breakdown of what you would like to use the funds for, and any associate costs. They want to see what you will do with the funds or if it was you just picking a number out of thin air.

Personal Financing

- Out-of-Pocket: Minimum of 25 to 30%
- Family/Friends

Equity Financing

- Trusted Business Partner with capital to contribute
- Investor

Other Source

Crowdfunding is a viable option depending on the type of business you would like to start. Most business owners who use crowdfunding to fund their businesses and are unsuccessful as it is typically because they assume that they can just design a campaign without promoting it. Like the internet, there are millions of people on crowdfunding sites, so how will people find you if you do not talk about it or promote your campaign.

Scale down/start small to where it fits your financial means is the most logical step if you cannot get outside funding for your business. There is nothing wrong with starting your business on a smaller scale so you don't incur any unnecessary expenses from the start. This also is a true benefit to people with minimal industry experience as starting small is less of a risk to obtain that experience. If an aspiring entrepreneur can make their concept work on a smaller scale, then they can grow into their ideal scenario instead of starting big and failing and most of the time and not being able to adjust to the failure or adapt to it.

Free grants

As mentioned at the beginning of this guide, although there are free grants out there on legitimate sites, very rarely if at all are these grants designed for the general public. Grants are made available because an economic development agency or governmental institution, whether it is a federal, state, or a local one has tasks within the community they need to solve and will occasionally recruit businesses to help them solve those economic issues. Thus, grants are offered as a way of bringing those specific businesses they need to solve those issues.

For example, a hair salon very rarely will qualify for a grant because most salons don't solve an economic need.

CHAPTER 6. BUSINESS STRUCTURES, LICENSES, AND OTHER LEGAL STUFF

Owning a food truck comes with its own set of laws and regulations. Specifically, I'm going to explain the areas of licensing and required permits. Some may find this process informative while for others, it can be a burden. I never said running a food truck was going to be all fun and games! All joking aside, getting your truck licensed and permitted is however an area where you'll need to pay particular attention.

While it may seem like a hassle just to get your food truck on the road, the required health codes and regulations have actually helped the food truck industry really take off and grow! By requiring trucks to pass all health requirements, it allows customers to gain confidence when it comes to consuming meals from food trucks. But requirements can differ because every city, county, and state have their own specific operating requirements.

Throughout the life of your food truck business, there are going to be some specific things that health inspectors will look for when they show up. This list is not comprehensive but it will give you a general idea of what you need to do to pass an inspection:

- Proof of ownership
- Use of an approved commissary
- Proper sanitary practices
- Proper food storage procedures

- Food maintained at proper temperatures
- Approved operating licenses

This is but a small list of items inspectors will check on in the operation of your business. During the course of your business, truck inspections will be conducted annually… and usually at random! So, it's good practice to keep everything according to code and to follow the rules. Any violations can bring on unexpected fines. And your business can be shut down if you have too many violations. The issues causing the shutdown must be corrected before you can re-open for business. Continued violations can lead to a permanent shutdown of your business. So, it's really important to develop good habits right from the very start so things don't get out of hand!

Just to reiterate what we just learned. You'll need to follow all vehicle requirements. You need to stay as sanitary as possible. Keep all foods at the proper temperatures. And maintain proper ventilation inside your truck. Violations can create a bad reputation that can destroy your business in an instant! I can't stress this enough!

Additional Certificates and Legal Documentation

In addition to obtaining the proper licenses and permits for your truck, there may be additional documentation you will need to file. Especially if you're operating your truck under a different name than what's listed on your business license. This can happen if you have an existing business but later open up a food truck under another name. If you are operating under a different business name, you would need a Doing Business As (or DBA) certificate. This is legally required if you're doing business under a fictitious name.

In most states, businesses need to register with the state tax agency and obtain certain tax permits as a seller. You'll also need an Employee Identification Number (or EIN) which allows you to identify your business on government forms and documents. It's exactly liked a social security number but this number is associated with your business. You'll most likely be assigned an employee identification number when you apply for your business license. An employee identification number is required whether you hire employees or not. If you have your business license but don't have an EIN, you can apply for one yourself at **IRS.gov**. The process is simple.

Protecting Your Personal Assets

When forming your business entity, you may want to consider incorporation. This will help protect your personal assets from liability. Incorporation helps you as the owner and as an individual. If someone gets injured because of your business, they'll sue your corporation instead of you and your personal assets. However, incorporating doesn't mean you're 100% protected but it does effectively separate you from your business and provide an additional layer of insulation from your personal assets. One of the most common methods of incorporating is forming a Limited Liability Company (or LLC). This option is great for small businesses.

As with any important business activity, you should consult with a lawyer first to weigh your options before you proceed. Once you have all the legal paperwork taken care of, it's time to explore where you can park your truck to conduct business. When it comes to parking your truck, you'll need to pay particular attention to city zoning and parking regulations. Areas of the city will be designated as commercial or non-commercial zones... In other words, you can't just park anywhere you choose. Visiting with the County Clerk will provide you with a good list of approved parking locations. If you play by the rules, you'll avoid parking violations... at least most of the time!

When you're out on the streets, it's a good idea to build relationships with parking enforcement officers. They may be a bit more lenient for accidental violations if they know you. Just be friendly with them! And definitely pay attention to local ordinances. For example, in a certain city, a food truck may not park within X number of blocks from a school during school hours. As with everything, there also some unwritten rules when it comes to food trucks. Brick-and-mortar restaurant owners may not want food trucks parking in front of their businesses and "stealing" their customers.

So be wary of your surroundings. Just use some common sense when finding a location. Some cities like Chicago, by law, don't allow food trucks to park within 200 feet of similar businesses. However, that could change. It's often a good idea to build relationships with non-competitive businesses and form a partnership that can be beneficial to both parties involved. This simple action can later translate to increased sales and customers you would not have gotten otherwise.

CHAPTER 7. PARKING

Next, you have to choose where you'd like to set up your truck. Certain areas attract food truck loyalists more than others and these are those places that you have to target. These are:

Famous Tourist Destinations

Why? Well, exactly because you know that people will be flocking around the area! There's a place in your town that people usually frequent and tourists from all over the world visit. If you can get a permit to set your truck up there, then you'd be on the right track. And, if this is the case, you might as well sell food that is connected to your area or food that your area is known for so you can be sure that people will try what you have to offer.

Malls or Shopping Districts

Again, there are a lot of people around these areas and everyone knows that shopping isn't for the faint of heart. Sometimes, people go from shop to shop and that's very tiring so, of course, they'd get tired and hungry then when the lines are too long in the restaurants at the mall, they'd look for somewhere else to eat – such as your food truck! This way, they can also bond with their family or friends more and have fun choosing orders from your menu!

Empty Lots

It's simple: when a lot is empty, there's a lot that you can do with it. Before a restaurant gets built or before people loiter around the area with nothing to do, why don't you get a license and set your truck up there? It's a good way to make money and attract people to try something new instead of just sitting around and doing nothing. Think of the empty lot as an empty mind — it's so open for possibilities and that's what you want your business to have! This way, it'll be easy for people to associate the lot with your food truck and they'll find it as a great place to hang around in.

Office Parking Lots

Working at an office is not always easy. There are times when the workload is just too much that people are forced to take back the food to their cubicle with them. In this case, they need to be able to eat somewhere nice and different, just to get away from the monotony of it all, but they also have to make sure that they won't burn a hole in their pockets. As for this, you may want to put up your food truck in an office parking lot so that office workers won't have to go far just to eat lunch or get themselves some snacks. As there are loads of office workers and long office hours, you can be sure that you'll definitely earn a lot! In fact, during lunchtime alone, you'll probably be able to get most of your capital back so this is definitely a good place for you to put your truck in.

Business Districts

Don't stick to one office alone — target the whole business district. This way, when people are out of their offices for an hour or so, they can just check out your food truck and eat something or buy something that they can take with them on the go!

College Campuses

You know how college kids want to try everything, right? So, of course, when they see a food truck around the campus or even just a couple of blocks away, they'll definitely be rushing to try what you have to offer — which is a good thing for your business. Plus, in this age of social media, they'll surely post photos of your truck and your products online which is a great form of free advertising for you!

Train/Bus Stations

These are places where people have to wait for their ride and more often than not, they'll be looking for something that they could munch on. So, when they see your food truck, they'll feel as if their prayers have been answered and they'll be thankful that you're there to save them from their misery!

Beaches

Not all people have the time to prepare food for beach trips. Sometimes, they just want to go there and of course, it would also be hard if they spend all their money on restaurants around the area as they may be too pricey. But, with your food truck around, they have an alternative to the usual fruits or kebobs, and surely, they'll be able to enjoy that.

Events or Festivals

Make yourself available and send your plans to event organizers. Chances are, they'll allow you to set up your truck in a certain festival or event because there are a lot of people around and of course, they wouldn't just spend their time listening to the bands, they'd also want something to eat — so you have to be able to give them what they want.

Sports Events

If it's an outdoor setting, great. But, if it's an indoor setting, that's okay, too. Just wait for the event to finish, or be there before it starts so

while people are waiting in line, they can order some food from you and they won't just stand there being bored.
Remember, before you park your truck in any of these spaces, make sure that you have the license and that you have talked to the right people so everything will be official and you won't have any problems.

CHAPTER 8. COMMISSARY KITCHENS

Photo by loustejskal.com on Foter.com

Did you know that there are strict laws and regulations regarding the overnight storage of your food truck? The most common facility for food truck owners to park their trucks is at a commissary. The commissary is the place where you are required to park your vehicle when not in use.

Commissaries can also be commercial kitchens. The reason why commissaries and commercial kitchens are required is that they help keep this industry in check and promote overall food safety. As a food truck owner, you will be spending a great deal of time working inside your truck. However, you're more likely to spend an even greater amount of time working in a commissary or commercial kitchen. The primary reason you need the services of a commissary is that it is illegal to prepare food you're going to sell from your home. Your food truck and commissary will need to be officially approved and needs to operate within the guidelines of local health codes.

Health inspectors will also check that you are using an approved commissary or commercial kitchen. Their main job is to make sure that your food is stored and handled safely whether it's inside your truck or at a commissary. Any violations can cost you time and money.

If you are already a restaurant owner, then you already have a commercial kitchen at your disposal. For those who don't, a commercial kitchen or commissary will need to be added to your list of expenses.

Costs of Commercial Kitchens

Commercial kitchens make their money by charging a monthly fee for you to use them but there are creative ways that can help you reduce your costs when it comes to renting a commissary. Commissaries usually charge a monthly fee with average costs that could run you from $800 to $1,200 per month. This rate varies highly because each commissary offers different types of services. If cost is an issue, a simple way to reduce expenses is to partner with and share a commercial kitchen space with another business. This way, you can split the cost of the monthly fees.

Some facilities can be bare-bones while others may include security cameras, round-the-clock security staff, electricity, fuel, or other necessary supplies. One of the requirements needed before renting a kitchen is liability insurance; however, each commercial kitchen or commissary will undoubtedly have additional requirements that are unique to each location. Just be sure to get all the specifics before signing a contract. If you have trouble finding an affordable commissary, there are some other options you should consider.

Commissary Alternatives

Local schools, churches, or other venues may have a certified commercial kitchen that you can rent. Additional options can include hospitals, firehouses, and even catering facilities that already house the types of equipment that you will need. If you decide to go with one of these alternatives, you will most likely have to coordinate the use times with the kitchen owners or other users. And it's important to understand that you may only have access to these facilities early in the morning or late at night. Depending on your menu and operation times, this might not be the best option for you. But with a little creativity, you can potentially uncover a great deal on a kitchen rental.

If you've located a commercial kitchen you want to rent, make sure everything is legal and the contract is in writing. That way, both parties know what to expect in case issues arise. Having your lawyer look over the contract is also advisable. The various rules surrounding commercial kitchens exist to protect the consumer and to promote safety for the food truck industry. You should never place your food truck business at risk by taking shortcuts when comes to cleanliness.

Services Offered by Commissaries

A commissary or commercial kitchen is essential for a food truck business to fully operate. They offer services that a food truck owner needs to follow regulations and laws. Not only you prepare food at a commissary, but there are a host of other daily activities that are performed there.

- ***Convenient access to supplies***

A well-equipped commissary will have the most essential supplies on hand that you can get access to in a pinch. So rather than having to rush to a store to purchase items that you need, you can easily get them at the commissary. The time savings can be huge when things are busy… and when are they not? However, you'll have to check on the fine print of your contract because there may be certain services you are required to pay for when you are based out of a particular facility.

- ***On-site storage***

One of the big benefits of a commissary is that you can get on-site storage of your ingredients and supplies. Commissaries are fully licensed and approved for commercial food storage. Depending on the amount of storage you have agreed upon at the facility, you can stock up and save money by buying some of your supplies in bulk without having to worry about where you're going to store it all.

- ***On-site parking***

Another benefit to a commissary is that you can park your food truck at the facility. It's not just a convenient place to park but it is also where you are legally required to store your truck at night. Remember you cannot park your food truck at your home.

- ***Charging stations***

During your service hours, your truck is most likely going to run off generator power. But when you park overnight, you will want to connect your vehicle to shore power to keep batteries charged or refrigeration equipment running. Your commissary should have adequate connections to keep your truck plugged in. The cost of the power could be included in your rent or may be charged separately for actual power usage.

- ***Cleaning and waste disposal***

Another important feature of a commissary is the cleaning and waste disposal facilities. After a food service, typically a truck will return to a commissary where cooking tools and equipment can be cleaned. There are strict rules regarding how often certain surfaces and equipment need to be cleaned. Your commissary will help you stay within regulations and avoid any violations.
Wastewater requires special disposal facilities because it contains grease and food particles that cannot be dumped into regular drains. Any used water and solid waste need to be disposed of at the facility. Heavy fines can be imposed if disposal regulations are not followed.

- ***Vehicle maintenance***

Time is money and it's never a good time when your truck needs maintenance work. However, many commissaries offer mechanics services on-site. If your commissary provides mechanical services, routine maintenance becomes a whole lot easier. Other services can include kitchen equipment repair, routine inspections, and other necessary work.

Your commissary is your partner as you work in the food truck industry. It is also a place where you can interact with other food truck owners who are working in close quarters with you daily. Not only are you part of the larger food truck community in your area, but you become part of a smaller more intimate group at the commissary. Sharing ideas and getting help is a lot easier when you've got the support group to back you up.

CHAPTER 9. PRESENTING OR BRANDING

In business, you must always test the market to see what will work. After you have done your market research and have decided on the elements of your food truck, you now have to decide what you are going to serve and how much you are going to charge clients (based on the complete market analysis that you have done for your business plan).

When you have already decided what you will offer, dive into the details. Let's start with the menu that you want to present. You must finalize the food that you are going to put on the menu. A very common mistake for food trucks is to have a menu that is more extensive than it needs to be, or that offers too diverse a selection. A large menu overwhelms guests, slows down ordering and processing times, and takes critical space in your truck for storage that could be used for your best sellers.

It is good to have professionally designed menu boards that fit your brand, and if your truck is wrapped or decorated to draw attention from diners, your menu should look and feel consistent with that work. If your budget does not allow this kind of branding on the truck itself, a chalkboard with very legible offerings works fine. In the beginning, this will give you the flexibility to alter your offerings based on actual sales, before a more permanent decision has been made.

Testing the market is a very important step in opening up your food truck business because you want to know what is likely to work before you actually go in and invest your time, money, and effort. There are several ways to test the market for your food:

Research on the Success of Food Trucks Like Yours

The first thing that you should do would be to research food similar to what you're making. If you see that their food is popular, you can bet that yours will also be popular if you make it taste great. The only challenge here would be how to market your brand and how to establish a name that distinguishes you sufficiently to avoid market confusion. This should likely have been done before reaching the test market phase.

Try a Formal or Informal Survey

Another method of testing the market would be to create surveys. We love the taste plate approach, where you invite friends and relatives to a party or gathering, and offer your proposed menu for their feedback. These must be guests that you absolutely trust to give you candid feedback. You really need to know if your items are too salty, or not spicy enough. This informal survey is very valuable to finalizing your menu. The exercise of making this food also gives you a feel for labor and preparation time, which will help you decide if it makes sense to offer a particular item at a certain price. With the internet, this is now extremely easy to do formal surveys, because you can make use of survey apps or programs like Survey Monkey or even Google Forms. You have to just decide on the number of respondents that you would like to have for your survey and their demographic (your target market). From there, you can send out your survey and adjust your decisions on the results. If the results show that people like your idea, you may not need to make changes.

"Dry" Runs

Before you spend a ton of money on a food truck, you should rent the truck you are proposing to buy, whether it is new or used, ask for a tester that will allow you to actually serve your food in real-time. A poorly located fryer or a fridge that is too small will be very hard to adjust once you have paid the asking price. You may want a different truck. Similarly, if the plan is to staff the truck with one or more additional employees, you need to get in the truck with them and see how you move together, to make decisions about who will perform what functions when the truck is in operations. Finally, this gives you a chance at a festival, event, or with a prospective lunch crowd to see if your food actually sells, and how much. It would never make sense to attack this business and spend the capital required for success without having tried these things.

Creating the Brand

Your brand to be effective must accomplish all of the following:

- It makes your business easily recognizable. The easier your business is to recognize; the more people will associate your business with a certain product or service. This in turn translates into more sales.
- It must be pleasing, fun, and a positive reflection on your offering. If you are all organic or locally sourced, your brand should reflect what makes you better.
- The brand should be professional, and consistent with your product. Some

brands do well with characters, or funny animation, a gourmet brand offering more expensive or exclusive fair, should be more serious. This is critical to avoiding customer confusion.

- Your brand should be highly visible, from the truck or otherwise, and unique to you, so that customers can see it and instantly recognize what you offer.
- Your brand is much more than a logo, or wrap on the truck, it is a feeling you are trying to share with the customer that draws them to your business. The logo is important, but all of your branded items, should be consistent with each other, and contribute to a greater understanding of your business and offerings.

All of the best brands are easy to remember and simple in design. The very best brands are the ones that you don't need to explain. Once a person sees the brand, he or she will immediately know what your food truck is all about.

The best brands also hold an emotional appeal to the target market. The emotional connection that you will make will depend on the demographic and psychographic profile of the target market. In other words, you have to know what makes them “tick”. You have to know what strikes their emotional strings and work your brand identity on that.

Lastly, your brand identity should be consistent. If you have multiple trucks, they should be similarly decked out with signage or wraps.

Examples of Cool Food Truck Brands

Just to give an idea, here are a few very well-known examples of food truck brands that made it pretty big in America.

In Boston, there’s a food truck known as “Roxy’s Grilled Cheese” which is very known for serving delicious grilled cheese sandwiches. They have different types of sandwiches with everything from guacamole to Applewood bacon. Of course, you can enjoy having a simple glorious grilled cheese with them as well.

If you are into slider burgers, Easy Slider in Dallas is the place to go to. For those who don’t know what sliders are, they are small bite-sized burgers. Sliders became quite a fad a while before the food truck industry took off.

When the food truck industry took off, slider burgers also followed. Easy Slider sells nothing but delicious slider burgers with Angus beef patties and many different kinds of other ingredients and spreads. These are simple, easy-to-share fun foods.

Another noteworthy food truck brand is Natedogs which can be found in Minneapolis. They sell classic hot dogs in buns with a twist. Their hot dogs have unique dressings like beer mustard, relish, and caramelized onions.

Exotic cuisine can be found in a food truck known as "Quiero Arepas" in Denver. The menu is inspired by the cuisine of Venezuela. They serve burrito dishes with shredded beef, mozzarella cheese, and black beans. Their wraps also have many different varieties of stuffing that can be mixed and matched to suit the tastes of the buyers.

CHAPTER 10. CREATING YOUR FOOD TRUCK MENU

Questions to Ask Yourself When Choosing Foods for the Menu

In order to plan the perfect menu, you should ask yourself some questions that will make the business of choosing which foods should go on the menu easy for you. Here are some of those questions:

- What's easy for you to cook? Can you cook hotdogs without burning either side? Can you flip pancakes like a pro? Do you know how to make delicious patties with just the right number of condiments? You have to determine what you can cook so you can narrow your choices down instead of overwhelming yourself with the thought that you should cook every dish in the world.
- What's your specialty? Of course, there are a couple of dishes that you know how to cook and that's exactly why you're planning to open a food truck business. But there will always be a dish that you're confident about and you can cook better than anyone else does. What is it? Think about it and think about how you can use it for this business. For example, you can cook Fettuccine Alfredo-like you're from Italy and you know that it tastes different from what others make. Think about that and

see if you can make more variations, or if you want to feature the said dish with some side dishes. This way, when people think about your food truck, they'll remember your specialty dish and they'd keep coming back for more.

- Which ingredients are easy to get around you? Maybe, you're planning to put up a hotdog food truck, but you're in an area where there are loads of fish and fresh products around. What do you do? Will you still get meat for the hotdog from another town, or will you make use of the ingredients close to you, especially if you can actually make great dishes out of them? Sometimes, it's important to look around what you can do with what you have around because that will save you a lot of money and may even make you closer to people around you, as well!
- What do the people around you love to eat? Or, what are they looking for? Get to know your customers. Of course, it may be impossible to meet every each one of them but it wouldn't be impossible to observe and make a

general assessment as to what kind of food they enjoy the most. This way, when you set up your food truck, you can be sure that at least one or two persons will try what you have to offer. On the other hand, you can also observe what's lacking in the area and you can check whether you can give them that or not. For example, New York is full of these pizza, pretzel, and hotdog kiosks and food trucks. However, there's a lack of sushi and ramen trucks or even trucks that maybe sell something organic. You see, there are so many things that you can cook and offer people so research on that. If you offer people what they're missing or what's not currently available in the area, you just might get a positive response because more often than not, people want to try what they still haven't before.

- What kinds of food can customers easily take with them? As a customer, it's important to know that you'll be able to eat something easy to bring especially because most people are on the go these days. So, you must take the packaging of your products efficiently so that people won't have a hard time with them.

- Which ingredients are too costly? Think about the dishes that you'll be making and see to it that you're not wasting too much money on ingredients, especially if you don't have enough budget, to begin with. Think about a dish that you can make and you know you're good at that won't cost too much. It's important not to waste a lot of money when you're only starting.
- Which ingredients are portable? There may be times when you lack ingredients in the truck and you have to buy some more from the nearest store—but what if it's a couple of miles away? You have to think about the ingredients that you'll be using, too, because they're important when it comes to the dishes that you'll be cooking.
- Which food products are easy to re-heat? If you're planning to set up an Industrial Catering Vehicle, it would be important to know which food products

can you easily re-heat without them losing their quality, and you have to learn which foods don't get spoiled easily, as well, as you'll be traveling around a lot.

- Will you focus on your expertise, or are you willing to try something new? Suppose you're famous for creating delicious and appetizing cupcakes. Are you going to sell them or make them the focus of your business? Or are you also willing to learn how to make other dishes and make use of them, too? Diversity is very important when it comes to food trucks, sure, but being confident with what you're doing is also one of the biggest keys to success.
- Will your menu always be your menu, or will you be able to change it? It's important to observe whether your customers like your menu or not and be open to changes if needed.
- What time will you be open and on which days? You have to create a schedule and you have to stick to it because when your customers notice that you're not around for a day or two,

and when they feel like you're not open at a certain given time, they may think that you're no longer in business or that you're not serious with what you're doing—and that's definitely something that you shouldn't allow to happen.

Guidelines

Then, when you finally decide what kind of menu you'd offer to your customers, you have to make sure that you get to cook the food correctly and that you ponder about some guiding principle that will aid you to create the flawless food truck meal for your regulars.

These guidelines are:

- You have to make sure that you are consistent. Consistent in what, you ask? Well, consistent when it comes to making good food. Remember that you're not planning to have people eat at your place and never come back anymore, right? So, you have to make sure that you always get to create good food, so when they recommend you to other people, they won't be embarrassed that they did so and you'd gain more customers, too.

- Make sure that you can cook in large quantities. Remember, you're not going to serve one person alone, so it's best if you learn how to cook lots of food in a short amount of time. But of course, it's not just quantity that is important, quality is also essential—and will always be essential, so make great food in large numbers and you're all set.
- Make food that you won't have a hard time serving. Food Trucks are mainly created for people who are on the go so you have to learn how to work fast but still make sure what you're doing is right. Create dishes that are easy to serve, so people won't be bored and there won't be more pressure on you.
- And, make food that would not spoil even if it's taken on the road. You have to expect that your customers will take their orders with them on the go. Of course, some people may stay at your food truck and eat but most of the customers may choose to just bring their orders with them. Take care of the packaging and make sure to use only the right kinds of ingredients.

Extras

As time goes by, you can also add more dishes to your menu and you may also add some other items in your truck that you could sell. These items include official merchandise with the name of your business, some souvenirs that customers can give away, and other things that will remind them of your business so that they won't forget it right away. Make sure though, that you leave them a good impression so they'd want to buy these extra items.

If you know how to plan your food truck menu, things will definitely be much easier for you!

CHAPTER 11. PURCHASING AND MANAGING SUPPLIES BASED ON CUSTOMER DEMAND

In order to cook great food, you need great ingredients. Where you get those ingredients can be a quest in itself! Thorough planning comes into play again for your ingredient list. Knowing what to purchase in advance can help you to become more cost-effective and efficient.

Start by listing out the ingredients that are necessary for you to operate your truck. An important factor in building this list is figuring out how much food you can keep fresh and stored safely. Look at how much storage capacity you have in your truck or in an off-site kitchen. As a general practice and to stay within safety guidelines, it's always better to run out of food than to sell food that has gone bad! Being able to judge the right quantities to purchase comes from experience. There are different variables in each truck so, it can take some time to figure out how much you need to buy to prevent waste.

Here are some of the most common factors that will help you determine the amount of food you'll need every time you open for business. The day of the week can have a large impact on the number of customers that will visit your truck. The time of day can have an equal impact on the number of customers you'll be serving. Certain events may require more or less food. Examples are large city festivals or private events. Your marketing efforts also have an impact on how much food you need to purchase because the better your marketing efforts, the more people will buy your food.

Where to Buy Ingredients?

So where can you go to buy your ingredients? Again, it varies between different regions and cities... but here are some suggestions. Most supplies can be ordered from wholesale distributors. But you can buy your food directly from manufacturers as well. Look to local suppliers too because they can often get you the freshest ingredients. Co-ops are also an excellent choice because you can combine your resources along with different businesses. This results in increased buying power that can save money.

Over the years, warehouse stores have become very popular for restaurant owners. Stores like Costco and Sam's Warehouse are the best examples of warehouse stores. In general, mobile food businesses can use the same wholesale companies and suppliers as restaurants. But if you're just starting out and don't know where to find food distributors, a good place to start is to search on the internet for your area. In Google, search for "wholesale food distributors" and then add your city after that search term.

Local Suppliers

If you're looking for local food suppliers, a good source of information is just to ask area restaurants. You'll find that most are willing to help out. A great place to buy fresh ingredients is from local farms, farmer's markets, and local fishermen. Sourcing food from farmer's market has become commonplace and it's actually a top selling point for today's food trucks. If you're using fresh, local ingredients, then you should emphasize that point as part of your marketing campaign.

Organic and healthy foods are always a crowd pleaser! Especially in current times because a growing number of consumers are more conscious about what they eat. However, using organic ingredients may result in higher costs that should be reflected in your prices. Research has found that health-minded people do not mind paying a little more for healthier, organic alternatives.

It's also a good idea to get to know your local farmers and growers. By building a relationship with them, you might be able to get some better or exclusive deals with them! In the end, you may need to use multiple suppliers to get all the ingredients on your list.

Saving Money with Co-ops

Similarly, co-ops are a popular option that can be a money saver. When you are part of a co-op, you'll be joining together with several similar businesses to buy in bulk. The result is lower costs for everyone involved. Partners in a co-op don't have to be other food trucks. You can invite similar businesses to join with you or find an existing group to be a part of. If you're unable to find co-op partners on your own, you might want to get referrals from distributors or farmers. They may have some names of other companies that are trying to do the same as you.

Warehouse Stores

If co-ops aren't for you, then another popular option is shopping clubs. These are better known as wholesale warehouse stores. Among these clubs are Costco Wholesale, Sam's Club, and BJ's Wholesale Club. If you're already a member of these clubs, you already know that there is an annual membership required to become a member and to purchase from the stores. As mentioned before a lot of restaurant owners buy from warehouse stores simply because it's so convenient and you can get large quantities at one time.

To summarize, you need good quality ingredients to prepare great food and there are several viable options for food truck owners. What works for one business may not be the best option for others. Again, experimenting with different food sources is the only way to know what is best for you. But it's not a set-it-and-forget-it task. In the course of your business, you'll most likely find that you'll change suppliers several times. That's just the name of the game!

CHAPTER 12. LAY OUT YOUR KITCHEN

Photo by Nicklas Lundqvist on Foter.com

While some people say that food is the only important thing in any food business, you know for a fact that it isn't true. Of course, it's also important for customers to be able to eat somewhere nice because no one really wants to eat in a truck that's rusty or that's not even designed at all. If you don't have time to set up your truck in such a way that it would attract people, it may also mean that you are not yet ready for this business and that you may have to think things through.

Anyway, there are some things that you have to keep in mind when it comes to designing and decorating your food truck. These things are the following.

The Theme

Suppose you're creating a burger business. It won't be right to use pastels as the theme or put photos of classic Hollywood stars on the walls of your truck, would it? You have to make sure that the theme you choose is connected to what you're serving so that your customers won't be confused.

Color Scheme

The main rule is to use the colors on the opposing sides of the color wheel. This way, everything will go together and your truck won't look like it's painted by a two-year-old. Also, it would be nice if the color scheme of your truck is also something you can use for the uniforms of you and your staff to make everything cohesive.

Seats

Some food trucks allow their customers to sit around the truck so if you can put out some chairs or anything that your customers can sit on, that would be good.

Utensils and Packaging

It would also be nice if you could set up the truck in such a way that your customers won't have a hard time getting the utensils that they need. Always keep condiments and tissues around because most customers need them, and make sure that you have environment-friendly bags that they can just pick up and put their orders in so they can take them on the go.

And of Course, Give It Some Life

The best thing that you can do with your truck is put some of your personality in it. This way, your truck won't be generic and when people see it, they'll be excited to eat. When people notice that a food truck has life and that it's something cool, chances are they'll go on and try your products—and that's something definitely good for you! Attract customers and they certainly will eat what you have prepared! Let your truck speak for itself.

What to Expect from a Custom Food Truck Builder?

Custom food truck builders can be found in many major cities. However, their clients can be local or national which helps expand your options when it comes to finding a builder for your project. Some of these companies ship custom food trucks to international locations as well.

Food truck builders usually offer full-service concept and design services for food truck owners. These builders' job is to offer their expertise to help you design the perfect catering trailer or truck. They are also knowledgeable in building vehicles that adhere to strict safety regulations. When employing a custom truck builder, you need to check out their previous projects and even talk to the owners of the trucks/trailers they've built to see if their customers are satisfied with the results.

Truck builders should be knowledgeable in safety and fire suppression systems as well as servicing electrical wiring and connections. They should also be familiar with the various types of mobile kitchen equipment as well as proper installation into a vehicle. Most custom truck builders should be able to take your project from start to finish depending on the services they offer. This is what you want to find because it's much easier to deal with one contact or company that can coordinate each phase of your vehicle build. This includes:

- Concept Planning and Design

- Electrical Systems
- Fire Systems
- Custom Detailing Services
- Graphic Design
- Plumbing
- Vehicle Wrap Installation
- Equipment Sales and Installation
- And More

If you are working with an out-of-state builder, confirm that they are familiar with the health codes and safety regulations for your city or state. Check to see if they've built vehicles that are operating in your state. In many cases, you may have to provide specific rules and information to your builder.

Completion times can vary widely between different builders with some stating that they can finish builds in as little as 2 weeks. However, you should always expect delays and extended build times. These delays can come in many forms including equipment permits, inspection scheduling, safety violations, equipment availability, and more.

During the planning and building stage, it's a good idea to imagine being inside the truck preparing meals to identify where the bottlenecks might occur. Arrange to personally inspect and evaluate the interior during the build phase. Every piece of equipment you put into your truck takes up valuable space. The last thing you want is to realize that you placed a grill or cooler unit in the wrong spot after your truck or trailer has been delivered to you.

Vehicle Wrap and Exterior Design

It's actually astonishing how much money can be spent building the interior of your food truck. Installing the necessary equipment with an efficient layout is a major part of a step van conversion. However, the exterior of your vehicle requires just as much thought and planning. With your mobile food business, you only get one chance to create a great first impression... This happens long before a customer has even ordered food from your truck. Within the first few seconds of seeing your food truck, potential customers will immediately make decisions of whether or not they will order food from you. They may buy from you at that moment or their first visit could come days or weeks later because they remembered your truck design.

Your vehicle wrap alone can single-handedly convey the overall tone and style of your mobile food business to your customers. Marketing experts have shown that vehicle wraps are noticed and remembered better in almost all types of advertising except for TV ads. In terms of expenses, designing and wrapping can cost somewhere between $3,000 and $5,000.

Your graphics should be bright and easy to read. The goal is to help bring attention to your truck and hopefully attract more customers. The vehicle wrap literally turns it into a mobile billboard that can express the atmosphere around your truck and give a sense of the type of food you serve as well as your contact information. Your wrap is one of the most important components of your brand. The graphics on your wrap can be incorporated into other elements of your marketing materials. This includes graphics for your website and packaging like cups, napkins, business cards, and more. According to some, one of the downsides of vehicle wraps is that the adhesives that are used on the vinyl panels only last about 5 years when exposed to the outdoor elements. However, that time frame is just an estimate and your wrap could stay intact much longer. If you operate in a metropolitan area with lots of high buildings, consider extending your wrap to the roof so that people looking down on the street can recognize your truck.

Regarding contact information on your vehicle wrap, be sure to also include your website address, Facebook, and Twitter tags on the sides and back panels of your food truck. The exterior without a doubt is the most visible part of your business so it's worth including your company details on the side. Keep your contact information large enough to be readable from a moderate distance but don't let it distract customers from your main logo and brand graphics.

If you want to save money on your exterior graphics, you don't have to wrap the entire vehicle. You can instead have just your logos printed on large vinyl sheets that act like giant stickers that your installer can adhere to the sides of your vehicle. If you want to coordinate the base color of your vehicle with your logo, consider having the exterior of your truck painted a single color before adding your vinyl logos and contact information.

Power Generators and Propane

When you're out on the streets, your truck needs to be self-sufficient. Every piece of equipment needs to be able to run on its own without a tether to hard-line power sources and gas resources. This is accomplished with propane tanks and electrical generators. Every food truck needs both to be able to operate unless you are at a food truck park or another venue with available shore power and gas hook-ups.

Portable generators supply the electricity to your onboard appliances which include refrigerators, toasters, waffle irons, blenders, payment systems, and more. Some trucks have compartments designed specifically to house power generators and keep them out of sight and reduce noise. Food trucks that do not built-in enclosures will have power generators placed next to their trucks with power lines connected through ports on the side of the vehicle.

Determining the size of a generator (wattage) you need depends on the equipment you have on board. The best way to do this is simply to take inventory of all the electrical appliances you will be using. Then add up the wattage or amperes (amps) required by all of your appliances. You'll also need to know which appliances you'll be using simultaneously to avoid overloading your generator during your preparation and food service.

Total wattage in an appliance is calculated using this equation:

Watts = Amps x Volts

For example, if your device is rated at 5 amps and 120 volts, the wattage requirements of that device are 600 watts (5A x 120V = 120 Watts).

Sudden power surges

After you've made your calculations, add up the total wattage to help you choose the right generator for your truck. To avoid potential shut-downs, you will need to anticipate some extra available wattage use when choosing a generator. Some appliances will draw more power on startup or at different intervals so you will need extra capacity for that. Fortunately, most generators have a surge or peak power rating. The peak power rating indicates the amount of additional wattage the generator can produce for very short periods.

A sudden draw of power may happen when you start any electrical appliance. For example, a refrigerator may require 2200 watts to start its compressor (starting wattage) and then 700 watts to run after that (running wattage). When you add up your wattage requirements, you need to use the starting wattage in your calculations.

It's a fact that running a generator creates a lot of noise not to mention the exhaust fumes. Most food truck owners will place their generators on the opposite side of their vehicle away from the service window to keep noise to a minimum for their customers. Some food truck owners may even use two generators at the same time. But be aware that you will need to haul these heavy machines in and out of your truck each time you use them unless they're mounted in a compartment on your vehicle.

Propane tanks

While you must know the electrical requirements of your appliances, you will also need to be able to anticipate the amount of propane you will need for your truck so you won't run out at the busiest time! Propane tanks have approximate BTU ratings that you will use along with the BTU ratings of your gas appliances to help determine how long a propane tank will supply gas to your truck. The BTU ratings on an appliance assume that you will be operating the appliance at 100%. For example, the calculations are made on gas grills with all burners set on high or a water heater set at the maximum temperature.

If you don't know the size of your tank, the first thing you need to do is measure the height and diameter of your propane tank. Do not include the height of the collar at the bottom (and top) of the tank. The collar is just used as a base to keep the tank upright. The diameter is measured at the widest point of the tank.

CHAPTER 13. HIRING AND TRAINING YOUR FOOD TRUCK TEAM

Hiring

When it is time to start hiring employees, write down what you want in your team. What are the essential qualities? What is the deal-breaking quality? Then stick to these. Always check references! It does help to get referrals from people you know, but it is not recommended to hire your friends and family if you can help it. Owning a business will change the way you see people. Entrepreneurs are driven and hard-working. You will quickly find out that most people are not like you. As disheartening as it is, you have to weed out the lazy, negative, and backstabbing people.

When gathering applications and resumes, screen them for spelling and grammar mistakes. Notice what the applicant looks like when they bring it in. Are they dressed well? Do they smell bad? Did they smile? These questions may seem ridiculous, but if they do not put any effort in when trying to get the job, imagine how lazy they will be when it comes to actual work. It is recommended to have another trusted person to interview with you. Another set of eyes and ears can only help you through the hiring process. If the job you are hiring for is easy to learn, hire for professionalism and attitude. Finding someone that looks you in the eye, speaks well, and is a happy person who will be much easier to work with than someone with experience but answers their phone during the interview. Imagine your customers interacting with this person.

If it feels like you are not finding the person you are looking for, keep looking! It might take time, but you do not want to make a bad hire in the beginning. This is the make-or-break time, and you need people you can trust.

Firing

Unfortunately, as much work as you put into hiring the right people, you will still find that people are not who you thought they were. It is important to try to train and document any counseling or talks that you have. This is so important, and even if you feel the employee is your friend, you need to document every disciplinary or corrective discussion. Keep an employee file and be honest with your employee. If you feel that their behavior is endangering their employment, let them know. Begin a performance improvement program where you meet weekly for 4 to 6 weeks to see if the behavior is improving. Let them know that if it isn't, they will be terminated. Usually, the behavior will quickly improve, or they will resign on their own. If not, you have ample documentation to show you were in the right to terminate.

When terminating employment, it is always a good idea to have a witness present. This protects both you and the employee. Disgruntled employees are known to make false claims, get angry, etc. A witness provides a layer of protection against lawsuits and any investigations that you may encounter regarding the firing.

Human Resources

There are many, many regulations that you must be prepared to follow when hiring employees. Some notices must be posted, policies you must have on hand, training that have to be given and documented, and more. Many of these are industry-specific. It is recommended to join your local association for your industry. If there isn't one locally, join a state or a national association. They usually have resources to help you stay in compliance.

Many times, your insurance agent can help you to know what policies you need to have. There will be many related to safety, non-discrimination, hiring and firing practices, etc. A company handbook is a must and should include all of your work rules and policies. Update it annually and have your staff sign that they received a copy and understand its contents.

Choose Your Vendors

When choosing vendors, it is a good idea to get references. Unless they are your only choice (for example, the water and sewer, electric, etc.), you can get other people's opinions of the services and products they offer. These days there are a lot of variables to consider. Do they have good prices? Good quality? How quickly can they deliver? Do they offer credit or only immediate payment?

In the past, you could be stuck using the only supplier in town. Now that there are so many delivery services and options to order online, you can negotiate to get the best prices. Buying local is still great but remember you must stay profitable! You are the only one responsible for the health of your business, so remember you must make hard choices sometimes to stay open. That could mean saving money and choosing another vendor.

CHAPTER 14. FOOD TRUCK MARKETING

Your marketing plan is technically a sector of your overall business plan but these days, it's perhaps the most essential part of your business. Without marketing, you're not going to go anywhere. Consumers have far too many options these days to frequent spots that don't work to grab their attention. The flip side is that spreading the word-of-mouth assessments is now easier than ever thanks to social media. Who knows? With the right ingredients, you might find yourself going viral.

Let's start with the basics. What is a marketing plan? Think of it as a business plan for your marketing. Just as how you'll refer to your business plan to keep tabs on yourself and use it as a reference document during tough times, your marketing plan functions the same way for your marketing efforts. In fact, your marketing plan is a more fluid document that will change a lot at first, and then more gradually as you zero in on the marketing strategies that work for you.

Your plan should serve as an instruction manual for all of your marketing efforts. Leave nothing up in the air. Detail everything meticulously. An important detail to document is your metrics. Every digital marketing outlet these days has a means of measuring your return on ad spend. With paid ads, it's very obviously displayed on ad dashboards. With organic methods, such as an Instagram account, metrics are more circuitous. You can track follower counts, new followers gained, and the number of people who tagged you, for example.

These metrics will help you figure out whether you're reaching your goals for customer growth and retention. Retention is an area that not enough business owners focus on. A returning customer is worth twice as much as a new one. This is why incentives such as a loyalty card are so powerful. They keep people coming back to you for more, and as your retained customers grow, they spread the word about you for free, leading to more new customers.

Your marketing plan will outline the strategies you'll use to promote new customer acquisition and growth. It'll also outline who will be in charge of these activities. What your budget will be and so on. Your marketing plan should complement your business plan. Indeed, you'll refer to your business plan quite a lot when you create your marketing plan.

Before we get into the details of drawing up a marketing plan, I want to caution you against dismissing marketing elements as corporate or management consultant mumbo-jumbo. Some of the jargon can indeed get needlessly complicated and markets like to coin ridiculous-

sounding terms, but they're important for your business. Marketing has evolved dramatically over the past decade and you need to remain up to speed with it.
So, pay attention to all the points as they'll ensure your business will be successful before you even launch it.

Competitor Analysis

If you live in an area that has a lot of food trucks, you're not going to have trouble finding competitors to analyze. The trick is to pick the right ones. Food trucks come and go, just like restaurants do. You need to pick the strongest of the bunch and reverse-engineer their business models until you know them almost as well as the owners do.

You'll need a spreadsheet to keep track of everything related to your competition. The most obvious data points are their menus and food styles. Note their prices and any loyalty programs they're offering. Go ahead and eat their food as well while you're at it. Keep a log of their locations over the past six months. If you live in an area that has cold weather, note how they managed that season. Perhaps they have a catering service that kept them afloat during that time? Or did they change their menu?

The easiest way to figure out their location is to take a look at the Instagram page. All food trucks have social media accounts with geolocation tags on them. This helps bring prospective customers in. It also makes it easy for you to figure out where the top locations are. Mind you, you don't have to copy their location patterns. These places will be extremely competitive, so they might not be the best choice for a newbie food truck operation.

Note how long they've been in business. Their website will list this. The longer a truck has been around, the better their reputation is and the more word of mouth advertising they have. In some ways, it can be counterproductive to study trucks that have been operating for a long time. They've built enough equity with their customers that they can take a few risks with locations and menu choices that a new business can't.

This is why it's better to have a spectrum of competitors. Choose one that has been around for a long time, one that has been around for a year or so, and another that has been recently launched and has been generating a lot of buzzes. You'll be able to find these new entrants through media mentions. Search Twitter for mentions of food truck names. The ones that show up the most are the most popular. Look at when their Instagram posts first began and you'll get a good idea of how long they've been around.

Here are the data points you need to collect, in addition to the ones already mentioned:

- Marketing channels.
- Target customer — You'll have to guess, but visiting the truck a few times and examining their feeds will give you a decent idea. You don't need a persona, just a few demographics.
- Average price point.

- Foot traffic at locations per hour.
- Estimated revenues.
- Advertising voice and style.
- Strengths.
- Weaknesses.
- Your advantages over them — Food quality is one but you want to look for something more than this.
- Their advantages over you — This is crucial. It'll prevent you from running away with yourself.

The objective of this analysis isn't to see if they're better than you. Just conduct it to get to know their business fully. Think of yourself as a business analyst whose job it is to figure stuff like this out. Rinse and repeat for all of your chosen competitors. Run it for at least three businesses, as mentioned earlier.

Note that this business analysis doesn't stop once you've completed your marketing plan. There are always new competitors on the scene and you need to stay up to speed. A deep-pocketed new entrant has the potential to blow you out of the water if you're not careful. So always stay up to speed with your market, even though it can be tough to do this when you're running a business.

However, this is what is needed to make this business a success. Most unsuccessful owners aren't willing to carry out these tasks. Doing this is what separates you from their ranks.

SWOT Analysis

SWOT stands for Strengths, Weaknesses, Opportunities, and Threats. It's a highly useful analysis to conduct since it forces you to be as objective as possible about your business prospects. Start by compiling a list of strengths associated with your business. Qualities such as your food, operations, ownership structure, and so on are valid business strengths. If your cuisine addresses a gap in the market, this is a strength as well.

If you're the only one serving a type of cuisine in the market, you need to take a long hard look at why this is. Don't naively assume that you're smarter than everyone else. For example, if you're a New Yorker and find that no one's serving Korean food at a particular location, it's probably because no one wants that style of food there. There are enough Korean food businesses in the city and the odds of you being the only one who's found this gap is close to zero.

Look for some of the tell-tale signs of a cuisine that's just not in demand. One of these signs is the presence of a decent-sized community that consumes this food but there is a lack of restaurants. This means, the community isn't big enough to support a business that cooks this type of food, and people outside the community aren't interested in it. Ideally, you want to see a small number of food trucks operating in the cuisine, but not too many.

If you find a large number of trucks operating, this doesn't mean you should give it up. Instead, look at how you can niche yourself within it. I provided examples of doing this previously in the book. You could be the dessert food truck or breakfast food truck, and so on. As a rule of thumb, try to avoid being a pioneer at anything. Food tastes are fickle. You might see huge demand at first as people are curious about the cuisine. However, you won't know how many truly like it until the next few weeks. Instead, it's better to operate in a niche that has demand and less competition. If competition is too high, niche yourself to reduce competition.

This is why your strengths and weaknesses tie with one another. It's very easy to assume a quality of yours is a strength when it's actually a weakness. Poor execution can make a strength a weakness as well, so make it a point to note what makes a quality a strength or a weakness. Food freshness can be a huge strength. Delivering this requires great process execution. Without it, it's a weakness if you're claiming fresh food but are delivering re-heated slop.

Be honest about your weaknesses. By "you are", I mean your business, not you. If you identify that marketing is a weakness, figure out how you can address this or mitigate it. Perhaps hiring someone to manage your social media is a good idea? It costs money but you'll earn it back through increased orders. To run a good business, you need to be able to delegate work properly. Many first-time business owners struggle with this. If delegation is your weakness as a manager, list it and work to mitigate it.

Next, it's time to list the opportunities your business has to stand out from the crowd. This is where your competitor analysis report will come in handy. Take a look at the gaps in their operations and execution. This spells an opportunity for you. Could you design an app that can deliver updates faster to your customers than merely social media? Investing in an app is a notable expenditure, but it does a ton to build a wonderful experience for your customer. You can collect a payment, the orders online, deliver customer loyalty points, and also inform people of your location without having to rely on just social media.

Your app can also function as a feedback collection machine. You can run contests and have people vote on what they'd like to see added. Best of all, you can create a community using your app. This is what consumers want the most, so give it to them. If no one else is doing this, or if they're not doing it well, there's your opportunity.

The final portion of your analysis is threats. Threats are items or occurrences that can disrupt your business. For example, bad weather is a threat, as is a lack of social media presence. Inability to execute your processes are prominent threats and can turn strengths into weaknesses. When listing process-related threats, make sure you get as specific as possible. For example, not having your lunch meal prep done by 11 AM is a threat. Many processes are a part of that, so break those threats down as well.

This gives you a clear roadmap of processes you need to execute to succeed. It makes it easy for you to refer back to them when you're confused or are unsure if your business is on the right path. As your business grows, the threats against it will evolve as well. So, make sure you're updating your analysis as often as necessary. Don't make the mistake of creating a document once and then forgetting about it.

CHAPTER 15. TAKE CARE OF YOUR FOOD TRUCK

If you take care of your truck, it will take care of you! Regular and preventative maintenance is the key to making sure your truck is operational when you need it. If your truck is sitting in a garage or commissary parking lot with engine or equipment failure, you will be losing out on income every hour it is not on the streets serving customers. Maintenance is important all year round but if you continue to operate during the winter months especially in harsh climates, it is even more important to stay on top of it.

Having a reliable mechanic is invaluable and the expense is well worth it if you want to be able to open for business whenever you want to. Customers can be disappointed when you can't show up at a location where they're expecting you. If you're handy, you may be able to do some of the maintenance yourself and save valuable time and money.

However, if you're not familiar with step van maintenance then having the number to a trusty mechanic is the best advice. The same goes for equipment inside your truck. A lot of repairs can be done on your own but there are probably parts of your equipment that are better left to the professionals.

It's good practice to have your mechanic inspect at the beginning of the season especially if your truck has seen little use over the winter. This will help ensure that your truck is in great shape and ready for the crowds!

Like the car that you drive every day, there are things you can check yourself to keep things running smoothly. Here are a few things you should add to your maintenance schedule that you can do on your own or with your mechanic.

Change Your Engine Oil Regularly

If you've owned a car or motorcycle, you already know the importance of changing the oil regularly. But how often you need to do it will depend on the type of vehicle you own. Step vans have different frequency requirements as well as using different types of oil than regular cars. The age of your truck will also need to be taken into consideration.

Your mechanic can help you figure out how often the engine oil needs to be changed in your food truck. Often, we delay changing the oil in our personal vehicles but don't delay the oil change on your food truck. Obviously keeping the engine properly lubricated will help it to last longer.

Maintain and Inspect Tires

Everyone has experienced a flat tire and the inconvenience it causes. But when you get a flat tire on your food truck, it can cost you income especially if you can't make it to your venue. A flat tire doesn't necessarily mean you will miss out on service but you will certainly be late… that is if you can get your tire repaired quickly.

You should inspect your tires regularly and monitor the air pressure. That way you know if there's a slow leak or problem and you can take care of it before you are in a critical situation. Driving with low air pressure can lead to reduced performance and handling on the road. It can also cause early tire wear. Of course, you can't predict when you might run over a nail or something but regular inspections can uncover foreign objects lodged in your tires. Check your air pressure before you hit the road as part of your daily checklist before heading out to a service. That way you can air up with a portable pump or stop at a gas station.

Checking the Engine Battery

The battery in your food truck is what gives the starter the power to turn the engine to start. You don't want to be faced with a dead battery just as you've loaded up your truck and ready to roll. You know how frustrating and inconvenient it can be to have a dead battery on your own vehicle! Like other parts of your food truck, your mechanic can perform some basic tests to determine if your battery needs to be replaced.

In cold weather, the battery has to work extra hard to crank the engine. But if you find that your battery dies once or twice, get it replaced! Vehicle batteries are relatively inexpensive when compared to the income loss due to a truck that won't start. The battery is something you could swap out yourself. If you need help, ask your mechanic to show you how to do it the next time around.

Check Your Fluids

The fluids inside your food truck's engine need to be monitored so they don't run low. They also need to be replaced or refilled at regular intervals to make sure your engine is running at optimum performance. As part of your

regular maintenance, schedule thorough vehicle checks at the beginning of spring and the winter. Also, perform periodic checks every couple of weeks to be sure nothing is leaking.
Fluids like transmission oil, antifreeze, brake, and power steering fluid are crucial to keeping your food truck running smoothly and without problems. Again, some of these are things you could change or refill yourself. If you need help, consult with your mechanic and have him or her show you how to do it.

Checking Belts and Hoses

One more thing to add to your maintenance checklist is a regular inspection of your belts and hoses. The heat of the engine coupled with constantly changing temperatures put a lot of strain on these parts. This heating up and cooling down can lead to excessive stretching or even breaking.
If you or your mechanic can see visible signs of wear on belts and hoses, it's best to change them out before problems happen. These types of repairs are fairly inexpensive and will keep your truck in excellent running condition.
Staying on top of maintenance needs to be one of the top items in your priority list among all

the other high-priority items that go along with running a gourmet food truck. One of the worst things that can happen on a business day is being stranded on the side of the road or stuck at your home base because of mechanical or equipment problems. Most of this can be preventable with periodic checks of the systems on your truck. Be sure to get tips from your mechanic so you can be a well-informed food truck owner.

It's also a good idea to share maintenance procedures with members of your staff so that there is more than one person that knows how to troubleshoot problems when they arise. Your food truck is the key piece of equipment to generating your income. Treat it with care and it will be a reliable partner as you continue to grow your business.

CHAPTER 16. DEALING WITH FOOD TRUCK HEALTH INSPECTIONS

If your food operation cannot pass a food inspection by the Health Department you will not serve food to the public and make money. Please keep in mind, not all of the items we will cover may apply for your state inspector; however, you will have a strong working knowledge of the expectations your business will need to achieve to prepare for inspection.

Purpose of Health Inspectors

A Health Inspector's sole purpose is to protect the public from food-borne illness. Hence, they are verifying that your business operation has good retail practices that exercise preventative measures to control pathogens, chemicals, and physical objects into foods prepared for consumption by the public.

Types of Inspections

Three types of inspections will be conducted by the Health Department:

- Pre-inspection before opening
- Periodic no-notice food inspections
- Temporary Event Inspections

A pre-inspection conducted by the Health Department before a food truck opening for operation and a periodic no-notice food inspection is the same type of inspection, with one exception. Periodic no-notice food inspections are used to observe you performing safe food handling practices while serving the public.

Temporary Event Inspections are food inspections conducted by the health department for events lasting 14-days in duration or less. Check with your county health department for the procedures required to serve food in this capacity.

What Inspectors Look For

Health Inspectors are inspecting your commercial kitchen or commissary, and your food truck for organization and process application. An inspector shared with me, that a Health Inspector can decide whether or not you have your stuff together by doing a 5-second observation of how you are organized. Specifically:

- Is your food permit properly displayed?
- Is your food truck clean?
- Do you know where your food thermometer is located?
- Do you know where your test strips for verifying sanitization water is made properly are located?
- Are you wearing an Apron and single-use gloves while handling food?

- Do you have a Wash (Green Bucket) and Sanitize (Red Bucket) filled with water?
- Is food stored on your truck 6 inches off the floor?

Most of these items are common sense, but you would be surprised how uncommon these things are to food truck operators that do not have a clue what they are doing. Here are the basic items of what inspectors look for.

Basic Inspection Guidelines

Commercial kitchen or commissary

1. **Handwashing sink**

- Hot/cold water dispenses
- Soap dispenser present & operational
- Paper towels present

2. **Toilet facilities**

- Self-closing restroom doors
- Paper towels present
- Soap dispenser present & operational

3. **Trash cans**

- Clean with trash bags
- Lids cover the can

4. **Warewashing**

- 3-compartment sink present

- Wash, Rinse, Sanitize
- Test strips are present

5. **Food contact surfaces**

- Surfaces are non-porous
- Cleaning solution available to clean surfaces

6. **Good food practices**

- No bare hand contact with ready to eat food
- Wash hands in hand sink, not in the dish-water sink
- Use single-use glove

7. **Cook foods to proper temp**

- 165°F Raw Poultry
- 155°F Ground Beef, Ground Pork, Shell Eggs
- 145°F Beef, Pork, Lamb, Seafood

8. **Chemical supplies**

- Use only approved food grade chemicals
- Never store food product together with a chemical product
- Label all chemical containers

9. **Store dry food goods 6 inches or more off the ground**

Food safety knowledge

1. **Time & temperature principles**

- 41°F - 135°F range where bacteria grow rapidly

2. **Metal stem thermometers**

- Must be food grade (NSF)
- Know how to calibrate

3. **Holding food for serving**

- Min hot holding temp 135°F
- Min cold holding temp 41°F

4. **Reheating food to serve**

- Reheat to 165°F
- Reheat rapidly 2hrs or less
- Reheat only once

5. **Employees that are ill can contaminate food, send them home if they are ill**

6. **Clean and sanitize all utensils and surfaces that touch food**

Know how to make sanitization water with bleach or QUAT (Ammonia). Food truck operators are required to have test strips available to ensure the proper ratio of water and bleach or QUAT solution is prepared correctly to effectively sanitize utensils and cookware.

Food truck operators also need to have proper cleaning buckets present to regularly clean food contact surfaces. Check with your local food equipment supplier for a Green Bucket and Red Bucket cleaning and sanitizer combo. The Green Bucket will contain a cleaning solution consistent with soap water and the Red Bucket will contain a sanitizer solution consistent with chlorine (bleach) or QUAT.

When cleaning food contact surfaces wipe down the surface with a cleaning rag submersed in the Green Bucket and sanitize the food contact surface with the same cleaning rag after it is submerged in the Red Bucket. After you have cleaned the food contact surface, keep the cleaning rag in the Red Bucket, which contains the sanitizer solution. Always, replenish these buckets with fresh water during a given food shift operation. I typically will change the water every 2 to 3 hours during a 10-hour shift.

CHAPTER 17. MISTAKES TO AVOID

Many people are jumping into the food truck industry, expecting to build a business that they can grow and profit from. It is undoubtedly a very exciting and high-profile type of business. New food truck owners are hitting the streets for the first time every month across the country. While the industry is still growing, it is an unfortunate reality that not every food truck owner is going to make a sustainable living at this. There aren't reliable statistics as to how many food truck businesses fail each year, but some experts estimate that the failure rate is upwards of around 60% during the first three years of business. This is about the same rate at which restaurants fail.

Before you start doubting the merits of the food truck industry, you should be aware of some of the factors surrounding those failures. That way, the warning signs can be recognized earlier, and preventative action can be taken. There are a lot of misconceptions or lack of knowledge for those who enter this industry. Many believe that just because they are a good cook means they can successfully run a food truck. While being able to create delicious tasting dishes is an integral part of a successful food truck, owners must realize that it is still a business that needs to be treated as a business. It's not necessarily a business you can invest in and just hire people to run it... at least not in the beginning. It takes a lot of hours and hard work to build up a customer base. A lot of growing pains are realized in the early years. And above all, costs need to be effectively managed.

Outsiders do not realize the amount of unseen work that goes into the daily operation of a food truck. There are many hours of preparation time like sourcing ingredients, location scouting, marketing, cooking, packaging, and more. In most cases, food truck owners are involved in every aspect of the daily routine and must be prepared and understand this fact. The time commitment alone is enough for some food truck owners to shut down in the first year. Anything worth building takes time to nurture and refine.

A great practice we can borrow from small to medium size restaurants is to keep your menu at a manageable size. Your truck has limited storage space as well as kitchen space. Too many items on your menu mean you need more space to store your raw ingredients. If your truck is theme-based, then only offer the most popular dishes for the style of food you are serving. That way, you can excel at a few dishes and maintain consistency. If you are cooking too many unique items, orders can get mixed up and confuse.

Making customers wait can lead to negative feelings toward your truck. Avoid offering dishes that take too long to prepare. Customers who visit food trucks generally expect relatively fast service, especially if they've been standing in a long line. In addition, the more people you can move through your line, the greater the profits. Having an efficient process from taking the customer's order to delivering the plated dish greatly benefits you and the customer.

Unexpected expenses can also kill an otherwise promising food truck operation. The daily costs of staying open can make it appear that money is flying out the door constantly. But when you have to pay for unplanned expenses, things really start to get tight. You're already paying for a commissary, propane, ingredients, serving supplies, staffing, and more! But often, vehicle maintenance gets overlooked. Your food truck will inevitably break down. Whether it is the truck itself or the equipment, any kind of repair is costly. To keep costs down, most food truck owners buy old trucks that are prone to mechanical failure. Some vehicle expenses can be manageable, but big breakdowns like transmission failure can severely put a dent in your revenue.

A lack of understanding of new technology can also contribute to the demise of a food truck owner. A mobile food entrepreneur needs to have a firm grasp of how social media like Twitter and Facebook can benefit a business. Social media platforms need to work in tandem with your traditional marketing efforts. Maintaining consistency in your cooking, locations, and customer service leads to happy customers that know what to expect each time they visit you.

Building and growing a reliable customer base takes time and constant nurturing. Having poor customer service can also turn new and existing customers away. A single case of poor treatment of a customer can lead to very damaging criticism on sites like Yelp and Urbanspoon. Customers who have never even tried your truck may brush you off without even sampling your food.

Like anything in life, you only get one chance to make a first impression, so make sure you have the right person interacting with your customers. Understanding where the points of failure can come from will better prepare you for these types of situations. Building a successful food truck is not a get-rich-quick type of business. Hard work, a sound business plan, and proper financial management are the solid foundation for any food truck entrepreneur. While it cannot guarantee success, it will give you an advantage over those that are not as well prepared!

CHAPTER 18. GROWING YOUR BUSINESS

The launchpad is ready to release and it is time to rev up the engines and stoves of the food truck. Marketing is essential to keep any business running. You should help the business to get noticed so that you can lure in customers. Competitors in the same field of business are never going to rest and make it easier for you. You must advertise and market yourself and your food product efficiently. Here are some marketing tips for the food truck business:

Set Up Weekly Specials

After the launch, you must gain speed and traffic in business. If a customer likes a specific food item like a Mexican taco, you could have "Taco Tuesdays" where you serve the customers tacos at half the normal price. This will spread the word and will assure you a lot of crowds.

Be One with The Community

Get close with the community you want to serve. Sponsor for a local sports event or try helping in a charity. Also, find ways to tie up with other business owners in the community.

Hold Contests

People love contests and they are an excellent idea to promote your food truck business. Promote contests through social media and other forms of advertising.

Celebrate Often

You do not need a big reason to celebrate. Opt for smaller holidays and make things exciting and new for the customers. Show the spirit of your celebration through the food you offer.

Have an Inner Circle

Treat your most valuable customers nicely and create an inner circle with them. Offer them discounts and earn their trust by being sweet and nice to them.

After all this, it is also important that you choose the perfect spot to put up the food truck. Make sure that you choose a place where there will be a lot of hungry people. Park your vehicle next to a commercial or industrial space. Also, make sure that there are no serious competitors around to spoil your day. When you want to choose a place, also find out about the events that might happen regularly at that place. Try to participate in such events and maximize your profit in doing so. Assure that you find out about the ease with which you can get the licenses to put up your food truck in these events. Do not feel bad to partner up. Partner up with a mall or building complex that will allow you to set up a spot on their property.

Tips to Sustain the Successful Run After Setting Up

It is essential to keep the business running in a smooth and controlled manner. This will make your brand profitable in the long run.

- ***Feel free to market yourself***

Marketing extends beyond the beginning phase and it is essential to keep the food truck running. Take advantage of digital media and its marketing platforms. Tweet about the places you are going to put up the stall, connect with Facebook, and maintain a Facebook page to post regular updates. Have a well-planned social media marketing scheme and try to lure in more customers by showing the merrier sides in dining with you. Also, make sure that you deliver the quality and service that you have advertised. False advertising can put a hole in the whole process.

- ***Think freely and do not attach yourself to an idea***

Even if you have found the perfect spot for business and even it had worked well for a long time, there is a possibility of dwindling sales. Take time to re-plan and think about moving to another new area. Do not be too rigid in the way you think. It is a waste of time and you might end up losing the business in the process.

- ***Expand on the revenue streams***

Change with time and try implementing new business ideas. Take risks and always be on the lookout for new opportunities. Cater to events and festivals to increase the profits you take. Get out of the comfort zone and try new and exciting things. Keep the energy and flow running.

- ***Be open to teaming up***

Do not feel bad about teaming up with other food truck owners out there. You could get a lot out of it because people who eat out of food trucks are most likely to change their trucks often. Pick a crowded place and a friendly food truck owner to club your business with. Cater to that crowded place together and get the best out of that situation. It need not be regularly but it is good to team up once in a while. People will also love the variety that you and your friend in business have to offer.

- ***Keep networking***

Make friends with people who have a strong influence over the place. Drop the prejudice and consider asking other truck owners to get valuable referrals for events and festivals. People might help you and you might even expand your network. Do not live in your world and miss out on the exposure that others have to offer to you.

- ***Make a good investment in your staff***

Make sure that you help the staff grow within their positions so that they stay trustworthy and faithful in the future. You must treat them with the respect they deserve and you must acknowledge their good work. The process of bringing in and training new staff is not only time-consuming but also costly.

- ***Put a good price tag on your food items***

Being new to the business doesn't mean that you have to offer food for a very cheap rate. If your food is tasty and has very good quality, feel free to charge the price that will benefit your system. It is vital to remember that people are ready to pay for the good stuff. Keep your eyes on the quality of the food you serve and you will automatically see business growth.

These tips and techniques are essential in your path to become a successful food truck owner. So, get out there and put out some interesting items on the menu to keep the hungry taste buds on fire. Serve with a bright smile on your face and complete love in your heart. There are a whole lot of people to feed in this world and it is high time that you realize that you can be the change you want to see. Thrive and work hard to serve the tastiest food on wheels and make sure that you touch the lives of people with what you do.

Food Safety

Food safety is a global issue, spanning several specific urban areas.

The food safety guidelines aim to prevent contamination of foods and that may cause food poisoning. It is done across various channels, some of which are:

- Sanitizing and proper cleaning of all surfaces, utensils, and equipment
- Maintaining a high standard of personal hygiene, in particular, handwashing

- Chilling, heating, and storing food correctly with regards to equipment, environment, and temperature
- Introducing effective methods of pest control
- Understanding food poisoning, food intolerance, and food allergies

Regardless of the reason you are handling food, whether it's part of your profession or cooking at home, it's important to always follow the proper food health principles. There are several possible food hazards in a food handling environment, many of which have severe implications with them.

According to the new annual study by OzFoodNet, Tracking the Instances and Causes of Diseases Potentially Transmitted by Food in Australia, 5.4 million cases of food-borne disease occur in Australia per year which are preventable. The incidents caused by these diseases are estimated at an astounding AUD 1.2 billion.

In American food businesses when referring to food safety, ownership is placed solely on the business itself. It must ensure that all foods handled and prepared within the business are safe to eat. Many are expected to hire a qualified Food Safety Manager to help the food business fulfill this duty.

CONCLUSION

A food truck business can be very lucrative because there are a lot of people who frequently eat at mobile restaurants. Rather than waiting for customers to get to your store,

you can go where they are and attract them with a special assortment of delicious dishes.
You can start and operate a food truck business with far fewer staff than it would need to operate a standard restaurant. This is also less demanding and requires lower operating costs compared to a conventional restaurant business.
You should start by having a clear business plan. In terms of the dishes served and the customers you want to attract, you need to pick the exact niche in the food industry. As most aspects of your business rely on these variables, you need to pick them from the very beginning. If you want to sell fast food, soups, pastries, ice creams, or multi-cooking meals, you need to learn.
You do need to know the age group you'd be targeting-whether teenagers, teens, college students, executives, or senior citizens. While the age ranges would overlap, you need to keep your target clients in mind before starting your business.
You have to keep in mind a particular target for your business. What's your business going to be for the next five or ten years? How many more trucks and workers would you have been

using by then? What kind of income do you expect to earn in the future? These are some of the goals you need to set very early on for your business.

Once you have a clear picture as to what you plan to do, you can obtain the necessary licenses and permits for your business. You need to be mindful that some towns and cities do not permit you to operate a food truck business. And you have to select your place of business based on the laws in force in the area.

If you have the permits, you need to buy your business a food truck. You may purchase a new or used vehicle, hire or even loan one for a certain amount of time. If you need financing for your business, you may need to find an appropriate bank or a private investor.

Once you have all of these in place, you can immediately start running your business. The secret of being successful in the mobile food business is being unique and offering something that no one else can offer. People still look for novelty and variety. You will become competitive in the Food Truck Business if you can deliver what they want.

Thank you for your support in this book. Best of luck in your food truck business.

CPSIA information can be obtained
at www.ICGtesting.com
Printed in the USA
BVHW070046270621
610451BV00003B/567